BARBED WIRE

BARBED WIRE

EVERYWHERE

By

K. E. PERRY

Published by Goss Publishing,
604 St. Paul, Kamloops, BC V2C 2K2 Canada
(250) 374-9472

Author & Illustrator Ken E. Perry
Photographs by Carol A. Perry
Editor Robert Goss

Canadian Cataloguing in Publication Data

Perry, Ken E. (Kenneth Edwin), 1943-
Barbed wire, barbed wire everywhere

Includes bibliographical references and index.

1. Frontier and pioneer life—West (U.S.) 2. Frontier and pioneer life—Canada, Western. 3. West (U.S.)—History. 4. Canada, Western—History. 5. Barbed wire—History. 6. Wire fencing—History. I. Goss, Robert B. (Robert Brian), 1956- II. Title.
F596.P47 1998 978 C98-910135-5

Every attempt has been made throughout the research process of this book to properly credit all individuals and companies for reference information which may have assisted in bringing this book to fruition. If you, or anyone you know, may have been inadvertently omitted, please accept our apologies and please contact us so that proper credit can be given in future editions.

First edition January , 1998
2nd Printing November 1998
Printed and bound in Canada

ISBN 1-894179-07-2

To my loving family and caring friends who were there every step of the way. Ever supporting and always patient of my barbed wire collecting activities, to them I dedicate this book.

Table of Contents

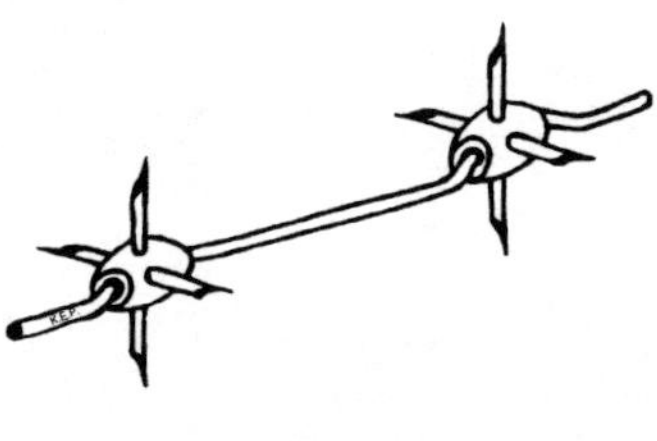

First Patent issued for
wire supplied with barbs.

Lucian B. Smith
Patented June 25, 1867
#66182, Kent, Ohio

Single Strand wire with
wood or steel cast spools,
spools rotate upon contact.

A PERIOD OF CHANGE

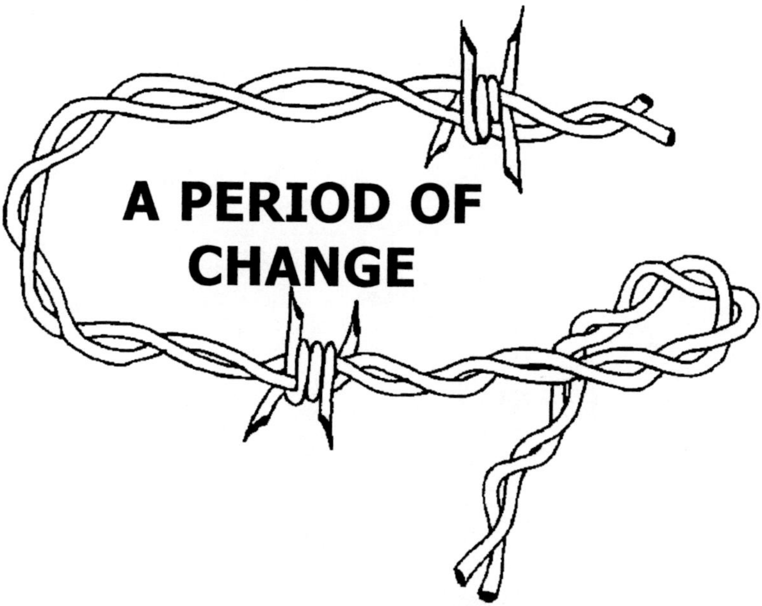

A PERIOD OF CHANGE
1859-1897

Since the dawn of the cattle industry of the mid 1800's, through the conflicts of the Civil War and onward to the Industrial Revolution, the life of the rancher, cowboy, farmer, and the mass settlement of the American West, changed very quickly during a relatively short period of time, approximately thirty eight years.

It was a period of immense change and opportunity for the ambitious. It was the period of the Western outlaw such as Billy the Kid, Jesse James, and John Wesley Hardin; and famous lawmen the likes of Bat Masterson and Wyatt Earp. It was a time of the trail-head cow towns like Dodge City and Abilene, Kansas, towns which were major railway shipping points to Eastern markets.

The West began at the Great Mississippi River in the east, across the central high plains to the Pacific Coast, and from the southern border of Canada to the Gulf of Mexico.

By 1861 the first transcontinental telegraph line was built. It ultimately brought an end to the Pony Express, a 2,000 mile mail express system which required a great deal of courage and determination on the part of hundreds of daring young men to get the mail through.

Western railways advanced very rapidly across the open range, opening the prairies to new markets. Native American tribal groups were no longer treated as sovereign nations. By 1871, under the Indian Appropriation Act, they were considered wards of the federal government, ending their traditional nomadic lifestyle.

By 1873, federal authorities estimate that three million buffalo were being slaughtered each year on the western prairies, almost bringing to extinction the great nomadic herds of

North America. Federal politicians believed that if you removed the buffalo, you could effectively eliminate Indian reliance on them as a prime food source, as well as ending traditional hunting practices that had been followed as long as they had occupied the land. Buffalo hunters rendezvoused for the last time in 1883, bringing to a close the large scale buffalo kill.

Five thousand miners intruded into the Black Hills of Dakota looking for gold, a direct violation of the Lakota treaty. The government solution to the invasion was to either lease the Black Hills or pay millions for outright purchase. This, of course, was not acceptable to the Lakota nation.

In a move to "encourage" the Lakota to accept the new terms, the Army was ordered to force the will of the government upon them. General George A. Custer was part of this engagement. With his command of over 500 men and officers he proceeded to the Little Big Horn valley in Montana. There, he split his troops into three groups, with one group on his left flank and the other on the right. He then took the third group and proceeded toward the Indian encampment along the Rosebud river. At the same time, more than six thousand Indians massed in the valley under the leadership of Chief Sitting Bull. After several fierce encounters his group was virtually overpowered. While seeking higher ground for a better advantage, he and his men finally succumbed to the superior force. After three days of fighting the battle was over and his command was almost totally annihilated, with only a handful of survivors left. This event is recorded in history as "Custer's Last Stand."

In retaliation for this disaster more troops were dispatched to the frontier to hunt down the various tribes involved and

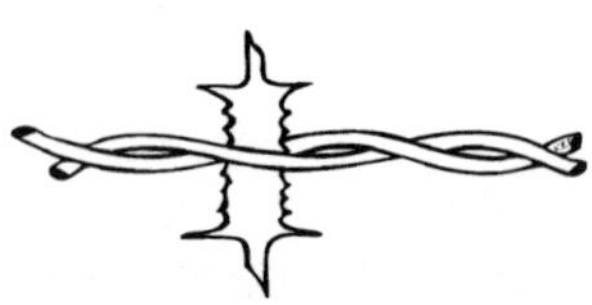

ROOSTER COMB
1895

force them back to the reservations. Not only was this Custer's last stand, but it was perhaps the last stand of a once proud and free Sioux Nation. Sitting Bull escaped the Army and crossed over the 49th parallel into Canada, where he lived for quite some time before returning to the reservation.

During this period, settlers purchased public land at twenty five cents per acre as granted by the Desert Land Act of 1877. The act allowed the purchase of 640 acres in arid regions, where more land is required to sustain large farming operations.

1883 saw Buffalo Bill (William) Cody, a one-time gold seeker, buffalo hunter, cattleman, fighter of Indians and Medal of Honor winner, promoting his famous travelling "Wild West Show." He pledged "a year's look into the history of the Wild West within three hours." It included Indians, mock raids on wagon trains, and even a special performance depicting "Custer's Last Stand." For a short time, Chief Sitting Bull was a special featured performer.

By the late 1800's the Industrial Revolution had begun, and machinery of all types were fast replacing older, slower methods of production. This process accelerated the life styles of the western people and the advancement of a nation.

The American automobile industry was created in 1896 by men such as Henry Ford, Ransom Olds, David Dunbar Buick and others. It was a time of great excitement, opportunity and wealth.

These events, and many more, all more or less began or ended during this short period.

This was also the era of the barbed wire fence, sometimes pronounced "Bob Whar." For centuries fencing of one type or another has played a major role in establishing boundaries to control man and livestock, as well as to mark the perimeter of public lands or private ownership of property. Wood, stone, hedges, bricks, steel and earth have all been used as fencing materials at one time or another. These materials were either too expensive to purchase, cumbersome to handle, ineffective by design, labour intensive to construct or to costly to maintain. Barbed wire has for more than a hundred years withstood the harsh elements of the environment. It will not rot or burn, is durable, reliable, light weight; and it is quick and easy to erect. It is relatively inexpensive to purchase, easy to transport and perhaps the only effective method for livestock control.

Fencing has been and perhaps will always remain the only deterrent to the keeping of something in or something out, and to mark the personal boundary of the land owner, government facility, public property or commercial enterprise.

The barbed wire fence was highly controversial. It was probably the primary element in creating the hostile division between the new settlers of the western plains and the cattle ranchers during the latter years of the 1860's through to the 1890's. It provoked a great deal of anger and resentment; it caused heated arguments, range wars, death and destruction of property the likes never before recorded. The debate on the use of barbed wire continues to this day and may never be completely resolved.

Out of all the basic frontier tools of the late 1860's and early 1900's, barbed wire became the main tool responsible for assisting the settlement of the West Central Plains, more so than the plow, rifle, or windmill. No tool helped further the establishment of land boundaries which, began the organized process of development, more than barbed wire. Without fences there would be no clearly defined boundaries, no method of demarcation. The laws of trespass and the legal ownership of

FOUR POINT
M.C. SHINN 1881

land would be in constant dispute. Survival of the fittest would be the order of the day and anarchy a way of life.

Barbed wire has evolved over the past one hundred and twenty-nine years into a reliable fencing material. It continues to meet the ever increasing demands and needs of land owners to this very day.

OPEN RANGE

OPEN RANGE

The first cattle brought to the open ranges of the Western frontier were Spanish longhorns, brought to North America some thirty years after Christopher Columbus arrived on North American soil in 1528. These longhorn cattle were tough and well suited to the harsh winters and hot summers of the western hemisphere. This was the beginning of the open range concept of raising cattle that prevailed throughout most of the great plains for a few hundred years, long before any real concern for fencing or land ownership. This practice continued right up to the late 1800's.

The open range was basically a wide open expanse of land with very few trees and no fences or other man-made barriers. In some regions it was a barren land of sand, rocks and the occasional mountain or rolling hill. Much of the open range of the mid-west was covered in bunch grass and sagebrush. It was an arid land with little annual precipitation, extremely hot summers and very cold winters.

Land was basically there for the taking and rarely did anyone complain. There were no trespass laws to worry about; days or weeks, if not months, would go by before anyone crossed the open range. If you did meet up with a stranger it was likely a cowboy looking for stray cattle which had wandered miles away from an adjacent herd.

Ranchers of the 1800's maintained very little land of their own, in most cases just enough to support their house, out buildings, corrals and gardens. However, they used thousands of acres of open range or otherwise public lands to graze their cattle on. Ranching on the open range was a very basic concept; cattle were allowed to roam wherever the grass took them. There were no fences and very few natural boundaries to keep them in a specific area, excepting of course the occasional river or mountain canyon. The grass was free and plentiful.

There was an understanding or unwritten law between ranchers, and that was to keep the open range unobstructed and open for the purpose of raising cattle and nothing else. However, change was once again on the horizon and this time it began when the Homestead Act of 1862 was established by U.S. President Abraham Lincoln. This Act essentially marked the beginning of western expansion. The Act provided the claimant with one-hundred and sixty acres of land. To qualify for deeded rights to the property, the claimant had to build a home, work the land and clearly demonstrate annual improvements.

Up until the close of the American Civil War in 1865, the common currency on the western plains was the trading or bartering of goods and services. The beginning of most business ventures generally started in gold mining, fur trapping, guiding, gambling, land speculation and anything else they set their mind to.

In almost every instance the successful cattle barons of the mid-1800's started their cattle empires with the successes of past ventures. After the American Civil war of 1861-1865, thousands of veterans from both sides took advantage of the free land. New immigrants from all parts of the world as well as settlers from the east came by the thousands to stake and claim land.

By the postwar reconstruction period, the building of a railway across the country from the east to the western shores of the Pacific was in full swing, a move that would unite the nation and transform the west. The Central Pacific and the Union Pacific were laying track faster than ever before. The more track laid, the more land the railway companies received from the federal government, 6,400 acres per mile of track laid.

The joining of both railways on May 10,1869 was celebrated at Promontory Point, some fifty miles northwest of Ogden, Utah.

This expansion met head-on with the open range culture of the cattle barons. Farming, small scale cattle and sheep operations soon took hold and the feud was on. Cattle rustling and fence cutting wars were not uncommon in the years that followed. Both sides fought for their beliefs (open range versus barbed wire fencing) and the profits it would bring them, but in fact the feud only brought death and destruction. Fencing became a basic requirement in the process of establishing boundaries, and the only deterrent for the protection of crops or the mixing of stock between one operation and another.

It was not long before the large cattle ranchers began to realize that they, too, would need more land if they were to survive. Ranchers began purchasing or controlling small and large tracts of land as fast as they could, and by any means available: pre-emption of land as granted by the homestead act; the purchase of land from smaller cattle operations and farmers; forcing the small operations out by intimidation and other methods of control; and negotiating long-term annual lease agreements for massive tracts of land from the federal or territorial governments. Land control and ownership took on new meaning and those who had assembled the most acreage with water rights would likely survive all others, and assure a profitable and secure future for generations.

The open range was fast coming to an end.

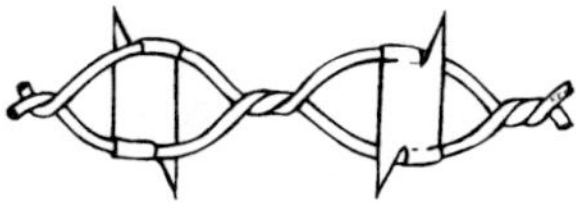

ELSEY PLATE
G. ELSEY 1882

FRONTIER TRAILS

FRONTIER TRAILS

Famous cattle and wagon trails such as the Santa Fe, Western, Goodnight - Loving, Oregon and many others were used to drive cattle from Texas to Colorado, Kansas, Oregon, California, Washington and Canada, to mention only a few. As well, they carried newcomers by wagon train to a new land which held promises of new beginnings, opportunity, adventure and disappointment; a land which was about to be transformed by settlement and hard work. North America's last frontier.

The huge cattle markets of the East were demanding a better quality of beef. The Texas longhorn was basically a lean and sinewy looking creature which roamed about aimlessly on the open ranges of the west. The meat was tough and fibrous, and rated less than adequate for the white cloth tables of Eastern restaurants. To improve the breed, controlled breeding programs and a system of range or pasture rotation were required.

Cattle entrepreneurs of the day imported cattle from England, Scotland and other parts of Europe to help improve the longhorn stocks for better meat production. These programs also required fencing and a lot of it. The only successful fencing material was barbed wire, as it was affordable, easy to transport, fast to install and very effective.

Other markets to the West and North brought cattle drives to military outposts such as Fort Bridger and Fort Laramie of Wyoming Territory and others which were scattered across the West; railways and telegraph work gangs; Indian reservations; frontier settlements and miners of the California gold rush; and, to a lesser extent, the miners in British Columbia (B.C.) and the Yukon of the Canadian North.

Canadian gold rush towns such as Barkerville of the B.C. Cariboo region, Manson Creek and Germansen Landing of the Omineca River region, the mining camps of the Big Bend country north of Revelstoke to mention only a few, were all part of a new growing market in those times. Between 1859-1870, cattle drives small and large totalling more than 22,000 head were trailed into B.C. from southern routes, due in large part to the discovery of gold.

In a very short time these famous trails came to a close. Barbed wire fencing and the advancement of Western settlement changed the landscape which ultimately ended the era of the open range. The West was on the move. Barbed wire was the primary tool that helped settle the central plains, and established law and order in a fast developing land of people from all walks of life.

ANVIL BARB
JACOB HAISH 1885

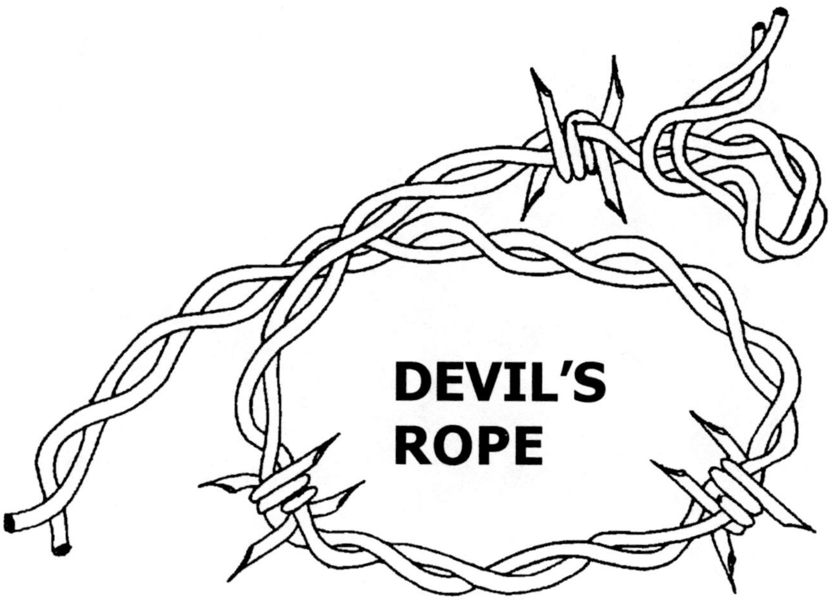
DEVIL'S
ROPE

DEVIL'S ROPE

Barbed wire has often been referred to as "the Devil's Rope;" it rips, tears and catches on almost everything it comes in contact with. Barbed wire is perhaps the most controversial fencing material ever made, excepting of course some of the extremely vicious kinds of wire used in military and prison applications. The debate on its use and applications continues to the present day.

Wire which was used for livestock control is basically divided into two categories, "Vicious" and "Humane". Vicious and Humane types of wire are very distinctive in appearance and were deliberately manufactured to meet certain criteria: one was the absolute control over livestock no matter what the consequences; the other was to bring an awareness of the presence of a fence to both livestock and man.

The former, of course, were designed to pierce the hide or flesh of the animal causing severe injury, and the latter to avoid injury. In any case the manufacturing industry struggled to meet the constantly changing demands of rancher and farmer alike.

Vicious wire types generally had long, sharp, fixed barbs which protruded at sharp angles in order to maximize its effect. Irreparable damage and sometimes death occurred from contact. The wire industry was in a quandary not knowing what type of wire would be required next. The future of barbed wire was somewhat doubtful.

LENOX TWIST
G.S. LENOX 1883

Humane wire or Obvious wire, the latter being more commonly referred to, was more in demand as the years passed. Some had spur like wheels (rowels) which moved or rotated upon contact. Others had special devices fixed to the main wire strands at regular intervals. These devices were known as signal or warning plates and were mostly made from sheet metal. In some cases small square wooden blocks were woven between the main wire strands. Cattle losses from injury or death were minimized.

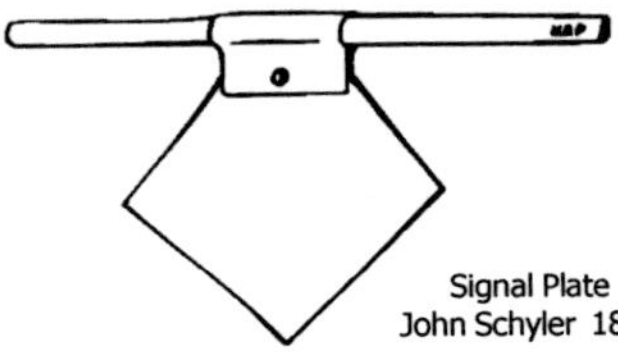

Signal Plate
John Schyler 1889

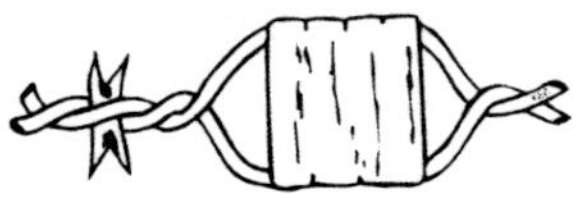

Wooden Block
H.B. Scutt 1880

Barbed wire was primarily made from steel and was often galvanized to protect it from rusting. Galvanization provided more strength as well as helped to keep the barbs from moving. Other materials used were aluminum, copper, and even stainless steel. Some types of wire had steel main lines with aluminum barbs and others a combination of copper and steel lines, while yet others had all aluminum main lines. Main lines came in a variety of shapes and sizes such as square, round, triangle, oval and were used in various combinations together.

Many other attempts were made over the years to manufacture wire with broad, flat metal barbs which served both as a barb and warning device. The attachment of warning devises was soon abandoned as the cattle became more aware or accustomed to the boundary effect of barbed wire.

BRINKERHOFF - MARTELLE
1888

Other efforts were made to develop alternate types of wire, such as the Brinkerhoff ribbon wire. It was very popular for a time. As strong and durable as this wire was, it too was soon abandoned because of the high cost of production.

As sales for Obvious types of wire dropped an industry standard began to surface, thereby securing a brighter future for barbed wire.

From all of the various kinds and styles of barbed wire manufactured in the early period, the Glidden "Winner" by Joseph F. Glidden, patented November 24th, 1874, survived all others and remains the most widely used wire pattern to this day.

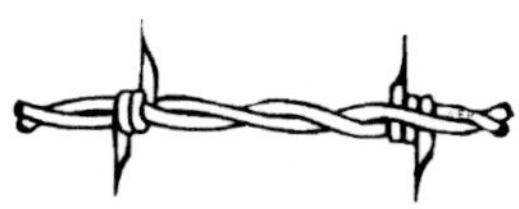

GLIDDEN WINNER
J.F. GLIDDEN 1874

DRIFT FENCES

DRIFT FENCES

Drift fences were generally constructed from east to west to keep cattle drifting from north to south. The drift fence was built like most other fences, with multiple strands of barbed wire strung between wooden posts. Post after post was placed in a straight line as far as the eye could see, spanning distances of two hundred miles across the open plains.

One of the earliest drift fences was built along the northern boundary of what was then called the "Cherokee Strip". The strip was approximately sixty miles wide and over two hundred miles long, an enormous corridor of land granted to the Cherokee Indians by the US Government for the purpose of going west to the buffalo hunting grounds. It was located in the north west part of "Indian Territory" (Oklahoma).

By 1886 the cattle industry was in trouble, as overgrazing had essentially exhausted the grasslands. Drift fences were erected to help preserve the grass lands during the winter by keeping the cattle from grazing too far south.

Then in 1885-86, and again the following year, were the worst winters ever recorded, winters that almost destroyed the entire cattle industry. Dead cattle by the thousands were found in the early spring along the entire length of fence. Cattle drifting south with the storms reached the drift fences and could go no further. Most either starved or froze to death. Cattle everywhere trying to find a way past the fence climbed on top of half dead or frozen carcasses to no avail. By early spring dead cattle by the thousands were literally piled up like cord wood. This period was commonly called the great or big "DIE-UP". The open range method of ranching was near its end.

Another notable drift fence, constructed across the southern central plains, stretched almost two hundred miles. It commenced from Indian Territory through the panhandle country of Texas, and ended west of the New Mexico border.

The attempt to protect the winter grasses of the open range from over grazing by cattle with the use of drift fences was soon realized as a failure and a detriment to the cattle industry in general. The use of drift fences was soon abandoned. Today remnants of the old drift fences can still be seen and are now considered as local heritage sites in most areas.

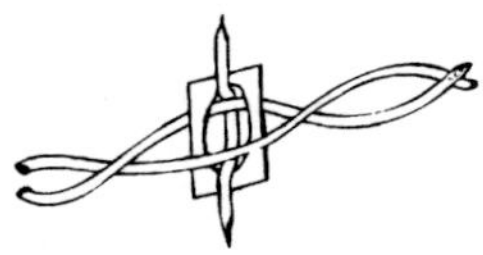

CLINCH WIRE
C.A. VOSBURGH 1876

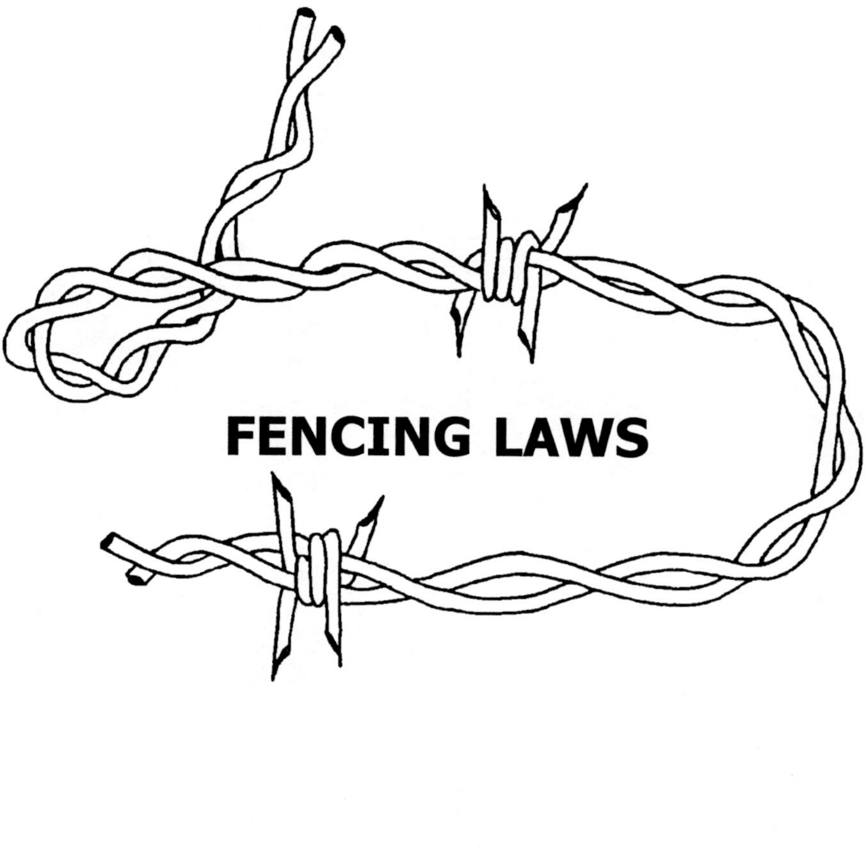

FENCING LAWS

FENCING LAWS

Between 1888 and 1894 legislation was passed which established new fencing laws for the control of livestock: the West took on the "FENCE OUT" law. This law was in essence an open range concept, making the land owner fence his holdings to keep free ranging cattle from crossing his land; the East, on the other hand, had taken on a "FENCE IN" law. This law brought into effect a law that made the rancher or farmer responsible for preventing their livestock from roaming at will.

The fence out law presented the advancing western railways with a great dilemma, that of fencing their right-of-ways to keep cattle off the tracks. Many thousands of miles of barbed wire were used along both sides of the right of way. This presented them with yet another problem, barbed wire thieves. This problem inspired new wire designs by the wire manufacturers, wires that could be easily identified and different to all other types. The wire had to be simple in design, but effective for the identification of thieves. The method developed was to make at least one of the strands of the main line square while the other(s) were round. Railway wire was in wide use throughout the Western Territories.

Anyone caught with railroad wire on their property or in their possession would surely be in for a great deal of trouble. The Western "FENCE OUT" law was eventually repealed and the "FENCE IN" law was adopted.

Railroad Spur Rowel
C.A. Hodge 1887

THE DRIVE AGAINST TIME

THE DRIVE AGAINST TIME

At the young age of nineteen, Charles Goodnight started a cattle business. Prior to the Civil War, his herd numbered less than two hundred. At the outbreak of the war he abandoned his herd to join the Texas Rangers. Upon his return to Texas at the end of the war, he found that Texas herds on the open range had expanded by almost six million head, and his herd by seven thousand.

Goodnight and his partner Oliver Loving trailed two thousand head of cattle north up the newly blazed trail named the Goodnight - Loving Trail to the mining camps in Colorado, to railroad construction gangs and to Indian agents. At this time, the cattle industry was driving thousands of head to the closest railhead cities such as Abilene and Dodge City to ship beef to eastern markets.

From the southern reaches of the Texas coast, famous cattle trails were heading north. Within a twenty year span six million cattle were herded up these trails; the drovers who trailed them to the railhead were called "Cowboys". They were usually young men in their early twenties; the pay was poor, the work hard, and sometimes they spent day and night in the saddle. After spending months in the saddle on a cattle drive, the rewards at the end of the drive were clean clothes, good meals, plenty of whisky and a wild time in town with the ladies of the local bawdy house. Although town business thrived on the uncontrolled spending habits of the cowboy, by the late 1880's the wild bouts of shouting, hollering and sporadic gunfire could no longer be tolerated without protest by the town folks.

The carrying of a revolver within town limits soon came to an end as gun control laws came into effect. Eventually the carrying of sidearms in most western towns and cities was no longer a necessity. Law and order was finally replacing the old frontier rowdy lawlessness. Barbed wire fencing was being erected across the open ranges at an alarming rate. Cattle drives to the north were regularly confronted by this foreign and intrusive barrier; it was becoming nearly impossible to negotiate the familiar trails which took them north.

By late August of 1881 the last of the major cattle drives to the northern railhead cities came to an end. Cattle herds were now well established in most parts of the central plains.

Barbed wire fences were everywhere, hardly a section of land was left untouched. The transformation of the west was for the most part complete. By 1896 the age of the Industrial Revolution had arrived. Trains, cars, and machinery of all kinds were now the driving engines of a new era that lay ahead.

The west was finally settled and the process of modernization was upon the land. Between 1877 and 1887 nearly five million people followed the trails west, settling the Great Plains and Pacific Coast regions of the American west. The west brought freedom for some, land for others, new beginnings and opportunities yet for others. All these ingredients helped form a nation.

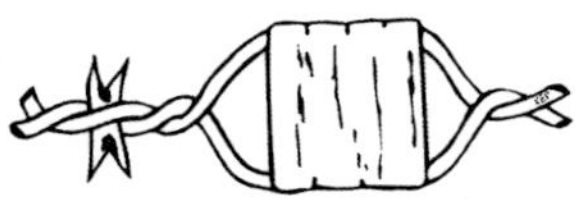

Wooden Block
H.B. Scutt 1880

HISTORICAL SITES IN WESTERN NORTH AMERICA

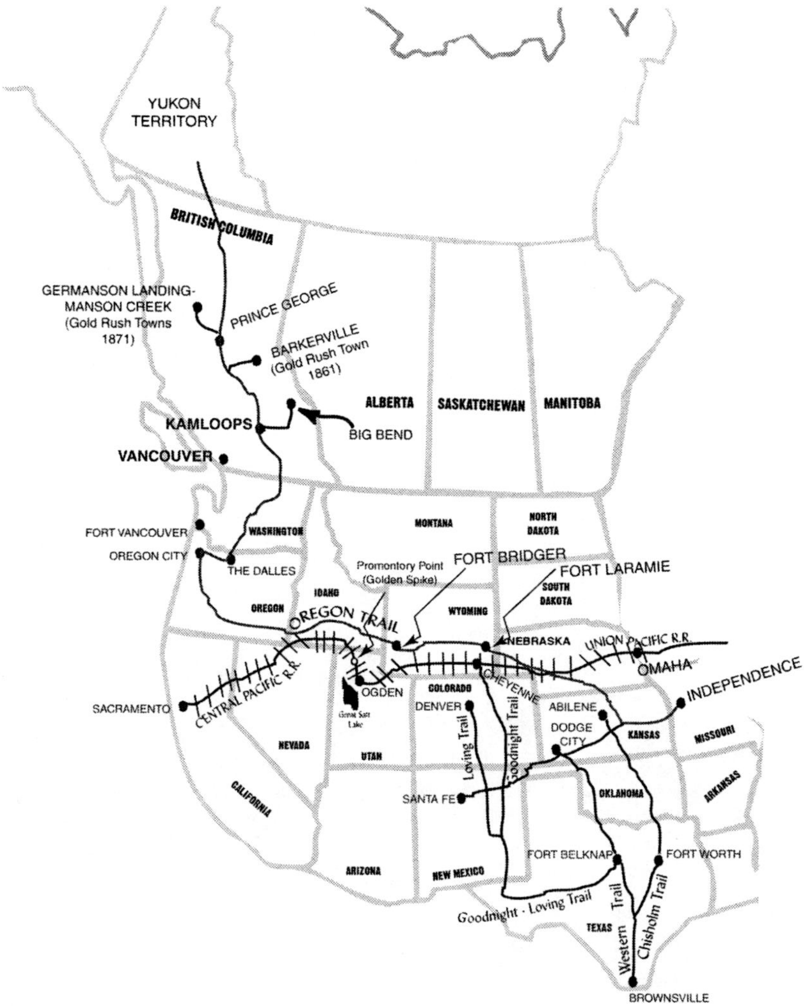

NOTE: Site reference points and travel routes are approximate and may not necessarily conform to scale.

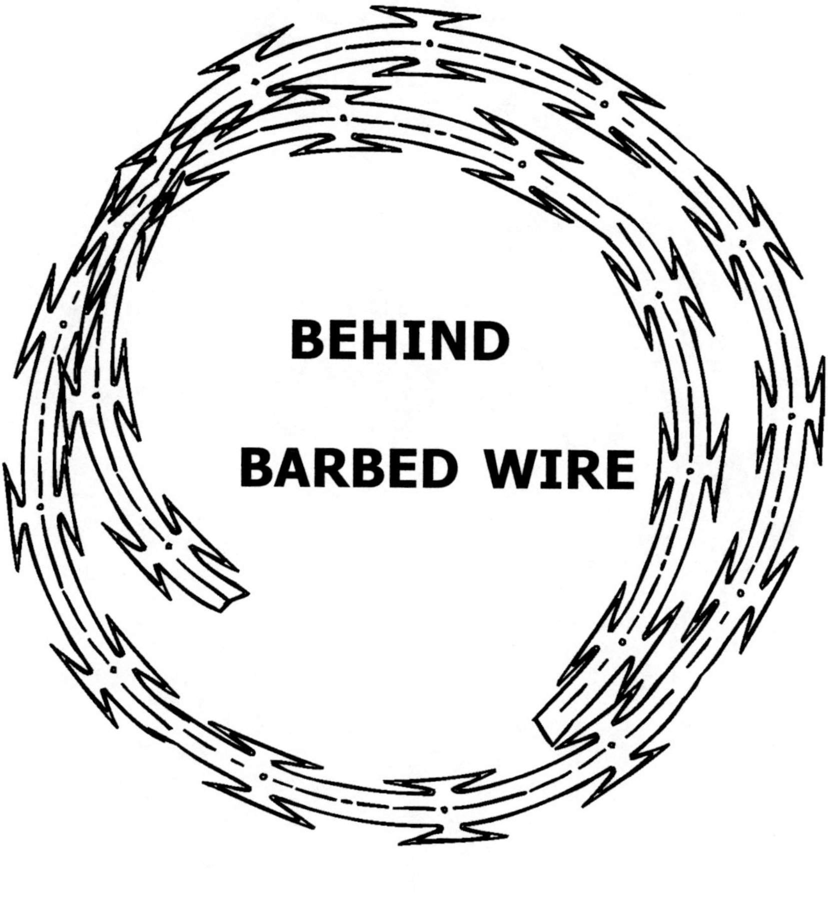

BEHIND BARBED WIRE

BEHIND BARBED WIRE

Barbed wire had other uses besides the containment of live stock or free ranging buffalo. One such use was in military applications. It has been used extensively around the world in almost every military engagement since the First World War to present day, including special applications such as military stockades for the containment of prisoners, foot entanglements and barriers for perimeter fences.

Foot entanglements were generally constructed at ankle height using a single or double strand barbed wire. It was strung in a quilt-like pattern, blanketing huge areas of the battlefield and making it difficult if not impossible for advancing troops.

Perimeter fences, on the other hand, employ barbed wire in such a way as to take on the appearance of a round concertina (accordion), tubelike structure. It comes in rolls of various diameters and when restraining straps are released, the concertina can then be deployed by pulling the multiple attached coils apart. The wire is then ready to be bounced into position along any designated line of defence, in single row, side by side double or triple rows, or stacked in layers row upon row to form a barrier while allowing a fairly clear field of view. This type of perimeter fencing is easy to erect and difficult to penetrate by man. Concertina wire is also used on top of fences, buildings and anywhere else barriers are needed.

Other types of concertina wire have been developed and are more commonly used today. The modern versions are called Concertina Barbed Tapes (razor wire). These new-style razor wires are used as a psychological and physical deterrent. It is more vicious and difficult to tamper with, provides exceptional perimeter security and extremely light weight, and is relatively easy to install. This new wire is manufactured from a continuous thin band of steel, machine-punched to form a vicious barbed pattern and is available in stainless steel. Some new wire designs by the military employ an explosive device which blows the barbed element up and creates shrapnel.

Barbed wire entanglements are generally deployed in field fortifications, beach obstacles for coastal defence, and trench warfare. It is an effective tool in the military arsenal.

Military wire basically evolved from agriculture barbed wire. It was modified with longer and fiercer spines (barbs) and made of thicker wire, higher tensile steel more resistant to cutting.

As the advancement of new technologies in materials, designs and applications are introduced, outdated military barbed wire becomes surplus and makes its way onto the civilian markets. Surplus military wire is often found on ranches and farms as well as in commercial and non-commercial security applications.

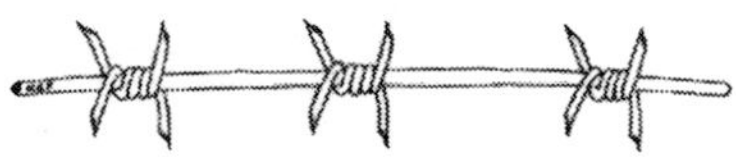

North American Concertina
War Entanglement Wire
(barbs 2 1/2" apart)

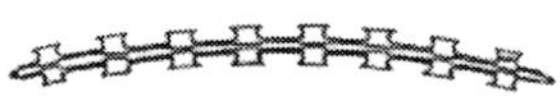

Concertina Barbed
Tape Wire
(Military Entanglement Wire)

Concertina Barbed
Tape Wire
(Razor Wire)

Thin Ribbon War Wire
(used in Vietnam war)

FENCE WARS

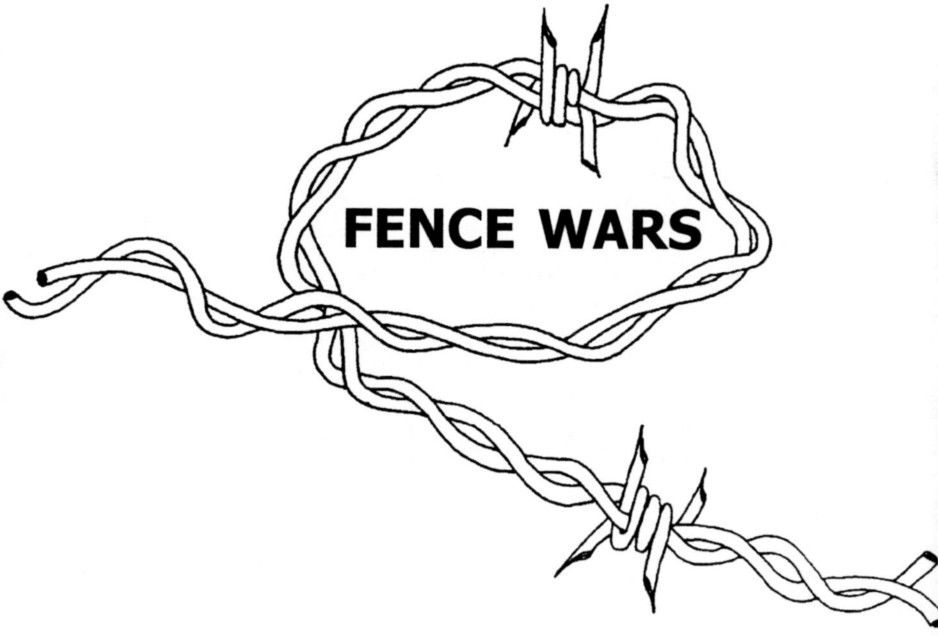

FENCE WARS

It was not uncommon for homesteaders (farmers) in the 1870's to shoot or kill cattle which wandered on to their lands destroying crops and gardens. Violence between rancher and farmer often ensued.

Soon after the invention of barbed wire, sporadic conflicts between cattlemen and nesters (farmers) broke out in the west central plains. Tempers were high and there was no compromise in sight. Fence cutting was as common a practice as the killing of cattle, and sometimes rustling. Both sides accused each other, thereby escalating the feud even more.

Many other acts of violence were associated with barbed wire during the late 1870's through 1881, especially in Texas. During the worst drought years Texas ever experienced, water and grass were difficult to find. The illegal fencing of public and private lands, including roadways, expanded uncontrollably in a last ditch effort to secure every bit of grass and drop of water for cattle. There was no regard for the farmers' legal right to be there; in fact, farmers were often killed, burnt out or run off. As these hard times continued the conflict reached warlike conditions. By 1883 half of Texas was reporting mass cutting of fences and the burning of range grass. Damages were estimated to exceed twenty million dollars.

By the end of 1883 the illegal fencing of public lands was no longer being tolerated. The continued practice of it was swiftly dealt with by the law. It was made a felony, punishable by up to five years in prison. Illegal fences were removed within six months and access to public roads was restored.

The 1892 Johnson County War in Wyoming is one of the most widely known public accounts of violence recorded on the western frontier. A total state of lawless conflict existed between the big cattle ranchers, small cattle owners, and farmers.

All parties engaged in cattle rustling and illegal branding of cattle, with all sides hopelessly locked in a violent effort to exercise their rights as they saw them. The rivalry and strife over allegations of cattle rustling became intolerable and finally reached the boiling point. With no apparent way to resolve their indifferences, the large cattle ranchers grouped together and hired outside gunmen, mainly from Texas. The cowmen were bound and determined to bring the entire weight of their organization down upon their arch enemies, no matter what it took. The small cattle owners, farmers and settlers were at the end of their rope.

A decade earlier these same cattle barons had taken the law into their own hands, and in a desperate move shot, killed, and hung suspected cattle rustlers and fence cutters. This move was meant to demonstrate to the state of Wyoming that the cattlemen were in charge, when it came to running the cattle industry and the land supporting it. In the years since this bold move, all sides were deeply entrenched in a war of words, beatings, threats and intimidation. Cattle rustling, brand alterations, and fence-cutting were all more or less part of their daily routine. The loss of life by all factions was a tragedy of tremendous proportions.

Once again, in the struggle for superiority, the cowmen hired gunmen to bring an end to the long-standing conflict. Huge rewards of money were offered to each gunman for every suspected cattle rustler killed. However, when the killers arrived by train in Johnson County, the cowmen's enemies surrounded their would-be killers and confined them for eight days, just long enough for the Army to intervene. This intervening action effectively ended the war. However, hard feelings and deep resentment remained for years to come. In the long run, victory went to the settlers and blame went to the cowmen monopolists. On one side were the settlers, on the other side were the open range advocates, and squarely in between was the barbed wire fence, a product of western expansion and progress.

Barbed wire did not always cause violence, but everywhere it was erected the closure of the open range culture was brought even closer. A new era was on the horizon, and barbed wire fences were everywhere.

DELFF'S LEAF
ARNOLD DELFF 1893

BARBED WIRE MANUFACTURING

BARBED WIRE MANUFACTURING

Barbed wire manufacturing was generally centered around De Kalb, Illinois in the 1860's. Rose, Haish, Ellwood, Kennedy and Glidden were very instrumental in the manufacturing and development of barbed fencing in this region.

Many opportunities in the new market gave back yard inventors a chance to get in on the action. Fortunes were made and lost, patent rights were bought and sold to the highest bidder. Hundreds of designs were patented; some were successfully manufactured and others never made it past the patent issue. From 1867 through 1895 it is estimated that well over 700 design patents were issued for barbed wire.

The race to get ahead progressed at breakneck speed. The buying and selling of patents for the right to manufacture was the goal and the quest for a greater share of the market was the prize. All of it was seemingly worth the gamble. Patent litigation against others for infringement of design and prosecution of counterfeiters of popular wire types plugged the courtrooms for years.

Barbed wire counterfeiting was a major concern and cause for legal action. Groups of ranchers and farmers formed alliances in protest over high priced wire. Manufacturing counterfeit wire, otherwise known as "Moonshine Wire," was a quick source of great wealth over a short period of time. Moonshine wire was produced in blacksmith shops, ranchers' barns and in small barbed wire plants. Back yard activities made fortunes for the unprincipled risk takers. Most moonshine operations were well financed and highly organized.

The producers of counterfeit wire were soon out of business, when judgements against them were handed down by the courts ruling in favour of the patent holders. Back damages for infringements were assessed and caused the ruin of most moonshine operators. However, not all moonshiners lost their bid to continue; some argued vigorously in court that their new and improved versions of the so called foundation patents were in fact more of an original design and therefore not an infringement against the prosecuting parties' original patent. Occasionally, judgement was made in their favor.

Eventually the counterfeiting of barbed wire settled down, bringing a calm to the wire industry and a refocus to the task at hand, the production and sales of barbed wire to the western settlers of the new frontier.

Glidden's first barbing machine consisted of a modified hand-cranked coffee mill. It was a slow and arduous process, a process that nevertheless got him into the wire business. Shortly thereafter a barbed wire machine inventor named P.W. Vaughan joined forces with Glidden in a partnership that would help launch their business careers. It was an enterprise with tremendous profitability. With Glidden's wire patent and Vaughan's patented wire machine, they could produce wire faster than ever before.

By February 1886 W.H. Putnam received a patent for the first power operated machine for applying barbs. In the same year many more power driven barbed wire machines were patented and put into service.

Barbed wire inventors and machine inventors soon combined their efforts through partnership arrangements. They either jointly manufactured wire or had other companies manufacture and sell wire on their behalf under license.

It is estimated that between 1874 and 1950 barbed wire production and sales ranged from ten thousand pounds to almost a half a billion pounds of wire. Wire was sold by the rail car load.

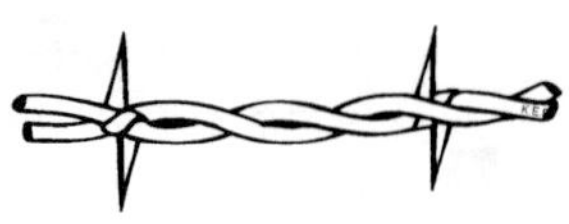

BAKER PERFECT
G.C. BAKER 1885

Elsey Plate
G. Elsey 1882

Barbed Wire Manufactured and Sold

1874..... 10,000 lbs.
1875..... 600,000 lbs.
1876..... 2,840,000 lbs.
1877..... 12,836,000 lbs.
1878..... 26,655,000 lbs.
1879..... 50,337,000 lbs.
1880..... 80,500,000 lbs.
1881..... 120,000,000 lbs
1882..... 160,000,000 lbs
1883..... 200,000,000 lbs.
1884..... 220,000,000 lbs.
1885..... 236,000,000 lbs.
1886..... 300,000,000 lbs.
1887..... 320,000,000 lbs.
1888..... 360,000,000 lbs.
1889..... 370,000,000 lbs.
1890..... 381,000,000 lbs.

Business boomed as the expansion of the West continued. Competition was keen and the fight for control of the wire market was extremely aggressive. Each company or inventor was trying to purchase all the patent rights they could get their hands on, or shares in companies making wire.

Of all the wire manufactured, there emerged only two wires that posted record production and sales. Leading the market was the Glidden "Winner," followed by the "Baker Perfect".

Wire production in North America began to decline just prior to 1950 as European manufacturers secured a foot hold in the North American markets with cheaper made wire. Almost half the market was imported by 1950.

To counteract these imports, American manufacturers were forced to change their ways and develop a better quality product at competitive prices. Through modern manufacturing techniques and other efficiencies, the gain on market share improved.

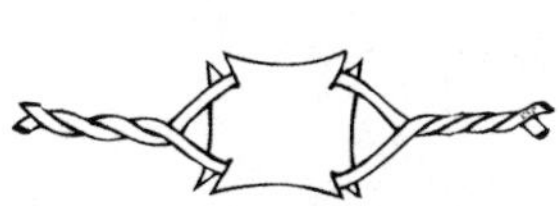

Stubbe Plate
John Stubbe 1885

PIONEERING NEW MARKETS IN THE 1800'S

PIONEERING NEW MARKETS IN THE 1800's

As the Eastern regions of the country began to use barbed wire, it became clear that the wire industry could not survive on the volumes that were being generated there. The eastern markets were slow to react to the new fencing material, due in large part to the availability and lower costs of other products such as wood, which was in abundant supply.

The real market for barbed wire lay to the West, the open range of the Great Western Plains; that part of the country where wood and other traditional materials were scarce. This new land was being transformed by settlement and a fast growing cattle industry, and was ready for the barbed wire fence, although no one was ready for the strife and division that came with it.

It was recognized early on by the wire industry that factory agents (salesmen) would be a valuable asset in the promotion of their products. One such factory agent was John W. Gates, otherwise known as "Bet-A-Million." Gates was hired by The Barb Fence Company (J.Glidden / I. Ellwood) which manufactured the famous barbed wire brand, Gliddens "Winner". Gates was sent to San Antonio, Texas in 1876 to sell barbed wire. Little did he know the opposition he was going to meet when he arrived. Ranchers were not ready to buy this new fencing material called barbed wire.

"Bet-A-Million," who would gamble on anything, was faced with a major dilemma: how was he going to convince these potential customers to buy his new product, a product which was proven in other parts of the country. The public was

BAKER PERFECT
G.C. BAKER 1883

unwilling to accept the new fencing concept. In a bold move he sought the permission of the town folks to stage a public demonstration. In the middle of town he built a corral made of post and barbed wire. Placing high-spirited longhorns inside, he then took all bets against the fence holding. With outriders galloping around the corral shooting guns and yelling loudly in an effort to stampede the cattle, the fence still held. Apart from a few wooden posts being broken, not one strand of barbed wire gave way, nor did any cattle escape. Gates could not write up orders fast enough, selling thousands of pounds of wire on the spot.

It was not long after this daring move that barbed wire became the fencing material of choice, and the way of the future. The word was out and the sales of barbed wire flourished throughout the Mid - West. "Bet-A-Million" Gates eventually became a millionaire and the chairman of the American Steel Wire Manufacturing Company.

ROOSTER COMB
1893

CUSTOMER TESTIMONIALS

A SUCCESSFUL TEN-YEAR RECORD.

Messrs. I. L. Ellwood & Co., Faribault, Minn., Aug. 16, 1886. The Glidden Steel Barbed Wire still "Holds the Fort" here. Farmers prefer it, and many will not buy any other even at a reduced price. During the last ten years I have sold hundreds of tons of it, and only in one instance was any fault found. Two spools were returned, having been found brittle. No other complaint was ever made, and all speak in the highest terms of its strength and durability. It stands our severe cold weather without breaking, a fact that cannot be said of any other wire. The Thick Set is especially prized, as no animal can go through a fence constructed of it. It is a great pleasure to sell goods that give such general satisfaction.

Yours truly,

J. C. N. COTTRELL.

WITHOUT A PEER.

Hamburg, Ia., August 15, 1886.

Messrs. I. L. Ellwood & Co.,—Gents: We have sold your wire since 1876 at Stuart and this place, and its sales have steadily increased in volume over any and all others. We have here among other customers, Messrs. O. S. Wing, T. Morgan, S. Finnell, T. Fuggett, A. J. Bartlet, and others too numerous to mention, who have Glidden fences made TEN YEARS AGO, all in as good condition, with barbs as perfect as now, proving it will wear a century. * * * * Many customers make unsolicited statements that they have moved the Glidden by dragging from one field to another over stumps, fence posts and frozen ground, without moving a single barb, while others will not even stand cattle rubbing against them without loosening the barbs. Such is our experience; and to-day the Glidden stands unrivaled, without a peer.

Very truly,

TOEDT BROS.

A GOOD WORD FROM THE "HILLS."

Deadwood, Dakota, Aug. 27, 1886.

Messrs. I. L. Ellwood & Co., De Kalb, Ill.—Gentlemen: We ave been selling the Glidden Steel Barb Fence Wire during he past year, and it gives by far the BEST satisfaction of any arb wire we have ever handled, and those who have tried it ill have no other, while heretofore we have had many comlaints and much dissatisfaction. Yours very truly,

AYRES & WARDMAN.

CHICAGO, ILL.,

BAKER WIRE CO., Chicago, Ill.

Gentlemen—We have constructed over 10,00 miles of barbed wire fence for railways, requiring over 8,000 tons of wire. For the past two years we have used in our contracts, exclusively, the "Baker Perfect," it having given the best satisfaction, and will build more fence to the ton than any other wire that we have used.

WESTERN FENCE CO.

MATADOR, TEXAS, March 3, 1891.

BAKER WIRE CO., Chicago, Ill.

Gentlemen—We are using "Baker Perfect" on our ranch and consider it the best and cheapest wire on the market. We have enclosed 1,100,000 acres of land on which are 100,000 head of cattle. We have built over 486 miles of wire fence, and consider we have saved many hundred dollars by using "Baker Perfect."

H. H. CAMPBELL,
Manager Matador Land & Cattle Co.

CLARENDON, TEXAS, Jan. 25, 1891.

BAKER WIRE CO., Chicago, Ill.

Gentlemen—I regard the Baker barb wire manufactured by your company to be superior to all other makes. I have inclosed 74,000 acres of land with your wire, and have on this pasture some 8,000 head of cattle, and I am not put to any expense by my cattle breaking the fence or by my neighbors' cattle breaking into my pasture. I also have a pasture of 1,280 acres of land inclosed with your wire, in which I have my buffalo and elk and moose and deer, and they have never succeeded in breaking my fence.

CHARLES GOODNIGHT,
Goodnight, Armstrong County, Tex.

WIRE COLLECTING

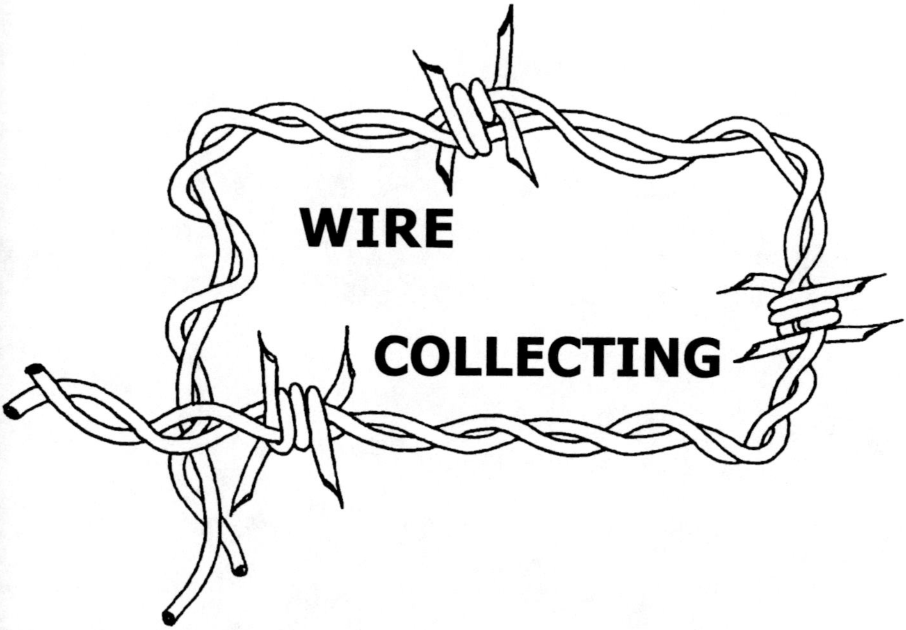

WIRE COLLECTING

The earliest known account of barbed wire collecting is from the 1950's, some eighty three years after its invention and introduction to the North American central plains.

As the awareness of the different designs of fence wire increased around the country, especially in the states of Texas, New Mexico, Kansas, Oklahoma and Colorado, there emerged an interest in collecting, and eventually the trading of wires began. During the 1960's and 1970's numerous articles, newsletters and books surfaced resulting in good, accurate history on the subject. One of the most widely used reference book on wires of this time, titled "Bobbed Wire Bible," remains to this day one of the most popular reference books for the collector. In the 1980's Annual Wire Shows developed rapidly across the nation, providing a venue for collectors to buy, sell, trade and display collections, but more importantly to establish new friends.

Barbed wire collecting has spread throughout the entire United States and to a lesser degree into Canada, Great Britain and Australia. It has grown into at least thirteen major collecting Societies and Associations.

Many museums across the land pay tribute to the barbed wire fence and its rich history associated with the settlement of the old West.

Two of the most notable all barbed wire museums in the country are the Devil's Rope Museum, established in McLean, Texas in 1991, and the Kansas Barbed Wire Museum in La Crosse, Kansas, which proudly boasts to being the World Capital for barbed wire.

SPUR ROWEL
C.A. HODGE 1887

The collector is often met with roars of laughter, strange looks of bewilderment or statements such as, "You Collect What?" or "You Can't Be Serious." These remarks are a common reaction among the uninitiated, since few people realize or understand the significance of barbed wire and the role it played throughout the 1800's and early 1900's.

However, within ten or fifteen minutes the uneducated becomes educated and more appreciative, aware, and understanding towards barbed wire and the influence it had on early settlement of the West.

The hobby of collecting wire is enjoyed by men, women and children of all ages, professions and ethnic origins. People are actively engaged in discussions on collecting, trading, selling or exhibiting their wire collections. It is a great social activity carried on throughout the country and around the world.

Wire displays can be seen at country fairs, cowboy festivals, museums, antique shops, special events national and international barbed wire conventions and shows.

Starting a barbed wire collection is easy and enjoyable. It will provide you with many hours of enjoyment. As in most hobbies, there are a few basic rules or guidelines, and in the case of wire collecting the following guidelines are generally accepted as the more important ones:

* Purchase some good reference books.

* Get acquainted with other collectors.

* Join a local Barbed Wire Collectors Association or a local museum which may have a collectors club or group.

CRANDAL'S CHAMPION
E.M. CRANDAL 1887

* Learn as much as you can on the history of barbed wire.

* Check barbed wire publications (news letters) for wire shows that you can attend.

* Buy a good pair of wire cutters and a pair of leather gloves.

* Get off the main roads

* Look for the "Unusual"

* Search out old farm and ranch sites.

* Get permission from the land owner before cutting a sample or going on his property.

* Never destroy property, private or public.

* Cut wires at least 18" long (standard length) shorter lengths has a lesser value.

* Always cut wire with barbs spaced evenly from each end.

* Discard broken or bad barbs or rusted through line wires.

* Try to get at least two (2) pieces of each type of wire, one to display and the other to trade or sell.

* Search out your find in the reference books and number your wire.

* Look up the number in the universal "Price Guide" for its estimated value.

* Set up a catalogue binder or an other suitable system to record your collection noting pertinent information on each find.

* Be positive of the wire identification. If you are not sure get a second opinion.

* Set up wire display boards or panels.

MERRILL'S FOUR POINT
L&J MERRILL 1871

* Trade wire through local events, club meetings or by mail.

* After your collection has grown to a good size, you may want to purchase some rare wire from a wire dealer or other collector to supplement your collection to enhance its value.

* When trading wires by mail never miss - represent what your are trading.

* Always give the other collector the option of returning your wire within a reasonable length of time if he or she is not completely satisfied with it, also reserve the same right for yourself.

* Always be punctual with your correspondence, sending wire on trade or for sale.

* Remember these guidelines and wire collecting will not get a bad name, encourage others to start a wire collection, new collectors are always welcome in the hobby.

*Never knowingly associate yourself with counterfeit wire or their agents, if such wires are collected they must be prominently labelled for what they are - "REPRODUCTIONS"

CRANDAL'S "T" LINK
E.M. CRANDAL 1876

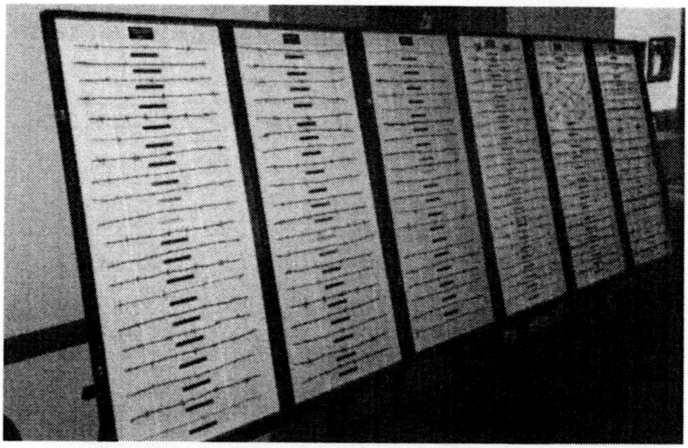

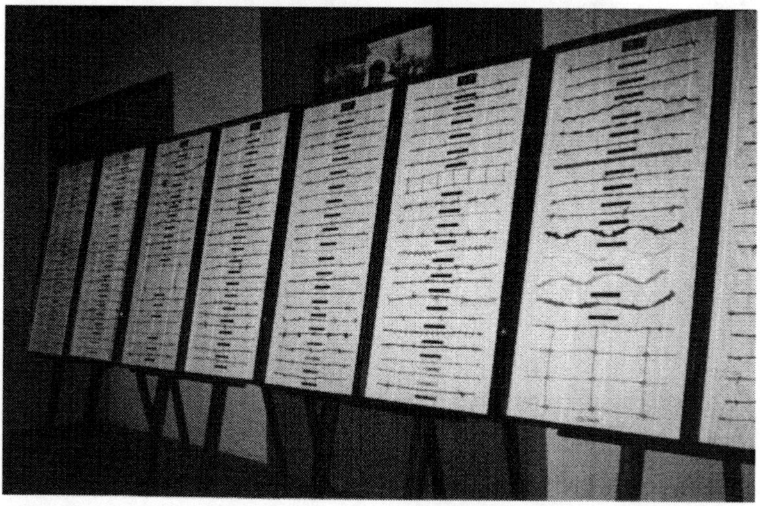

Wire Collection of Ken E. Perry - 1996, consisting of just over 300 pieces.

*Start Now!

GET "HOOKED" ON BARBED WIRE!

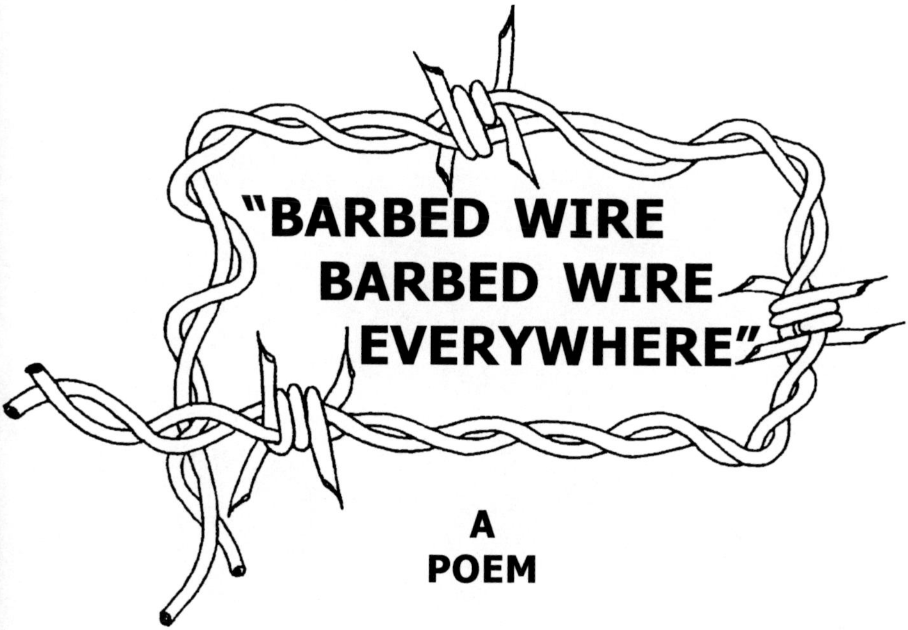

"BARBED WIRE BARBED WIRE EVERYWHERE"

A POEM

BARBED WIRE, BARBED WIRE EVERYWHERE

A Poem By K.E. Perry 28,4,1997

Barbed Wire, Barbed Wire everywhere.
It catches, it rips and even tears.
Can't get away from it no matter what you wear.
Barbed wire, Barbed Wire everywhere.

A long time ago way back in time,
When open range was in its prime.
You could wander here and over there,
Making your way without much care.

Civil War over to everyone's delight.
Western expansion was primed for a fight.

The land grab was on. Each man on the run,
Staking and claiming till that day was done.

Fences went up at a high rate of speed,
All types of materials a requirement indeed.

Constructed of stone and sometimes of wood,
Long hauls often necessary which was well understood.

Planting of shrubs and trees of the land,
Hardly worth effort, a dubious plan.

Then along came a man who brought irrevocable change.
Barbed wire the invention, Glidden his name.

The end of an era, cow trails of the past.
Men who pioneered them brought progress that would
last.

High in the saddle, proud as can be,
Remembered for their efforts for an eternity.

No longer these men, of vision and grit.
Post and barbed wire was the cause of it.

Cowmen and farmers unable to agree,
Which side of the fence the cows should be.
Cattle were shot for their drifting ways,
Gardens and crops ruined, now who's is going to pay?

Open range culture, a cowboys life.
Meets expansion head on and all its strife.
Re-settlement of Indians from east to west.
Advancement of foreigners, a fight at best.

Ownership of land was important to some,
Those who had it put the rest on the run.

The struggle ongoing, cowboy out for space,
All the nester really wanted was a spot for his place.

Fences went up, the boundaries were marked,
"Stay off this land!" the owners barked.

Tempers were high as range wars raged on,
Challenges by many that would never see dawn.

The source of the trouble was the cutting of fence,
Each side had their reasons which made little sense.

They pressed on regardless as arguments ensued.
Death and destruction the result of the feud.
Laws were enacted and peace finally came,
The reign of terror was now at last tamed.

Mile after mile of multiple strands,
Fixed to posts across the land.
Hundreds of miles of straight taut lines,
To stop drifting cattle and make easy their find.

Taming the west gave cause and concern,
For the cowboy's way of life was taking a turn.
Acceptance or not of these changing new ways.
Barbed Wire fencing was here to stay.

Anger and resentment were often the way,
Causing good men to be caught in the fray.

The end was near, open range fast closing,
Cowboy and horse no longer roaming.

The way was unclear, no trails to discern,
The laws of trespass were a major concern.

The barriers are up and here they must stay,
Some say it's good and some say nay.

Barbed Wire was with us, it's time had come,
Glidden and others had finally won.

Barbed Wire Barbed Wire EveryWhere.

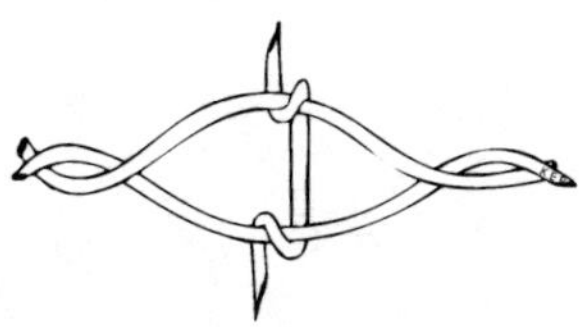

Decker Wide Spread
A.C. Decker 1884

EPILOGUE

EPILOGUE

For more than a hundred and twenty-nine years barbed wire has been the subject of controversy and turmoil. It was the direct cause of the fence cutting wars throughout the west in the 1870's through 1881 that resulted in a tremendous loss of life and the destruction of fences and other property worth millions of dollars. It was a prime source of anguish and frustration for many during those years.

Drift fences in the northern parts of Texas led the cattle industry almost into total ruination during the harsh winter of 1885-86 and again the following winter. Barbed wire was also responsible for the development of laws which brought order and an end to a wild and somewhat lawless frontier. It effectively closed the open range and ended a way of life enjoyed by the ranchers of the past. It ushered in a new era of land use management and control. Barbed wire fencing can be seen almost everywhere you go, and to this day remains the principal type of fencing material used in livestock control across the length and breadth of North America.

ANTIQUE BARBED WIRE

Antique Barbed Wire

Anvil Barb
Jacob Haish 1885

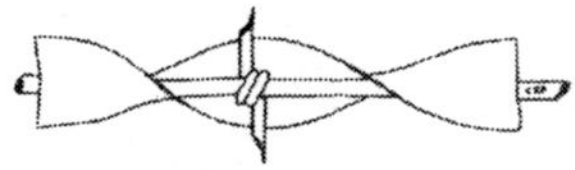

Brinkerhoff-Martelle
1885

Rooster Comb
1893

Merrill's Four Point
L & J Merrill 1871

Elsey Plate
G. Elsey 1882

Four Point
M.C. Shinn 1881

Lenox Twist
E.S. Lenox 1883

Crandal's "T" Link
E.M. Crandal 1876

Spur Rowel
C.A. Hodge 1887

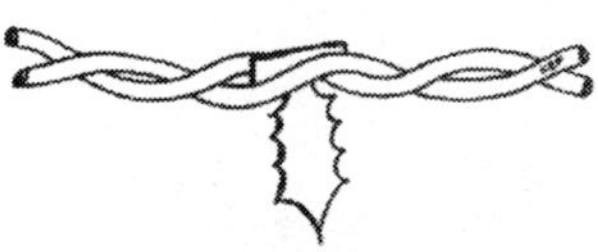

Delff's Leaf
Arnold Delff 1893

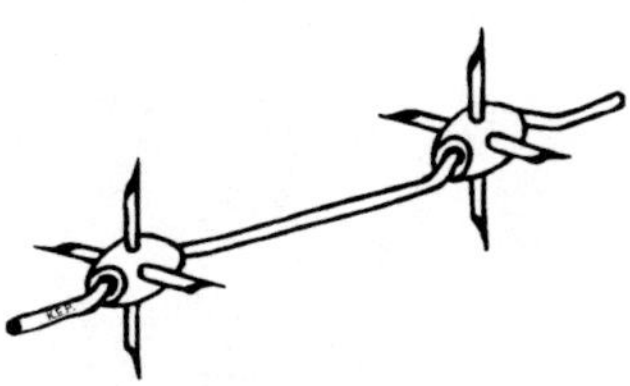

Lucien B. Smith
June 25, 1867

Railroad Spur Rowel
C.A. Hodge 1887

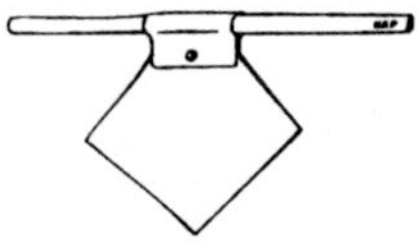

Signal Plate
John Schyler 1889

Reynold's Necktie
H. Reynolds 1878

Ribbon Wire
J. Brinkerhoff 1879

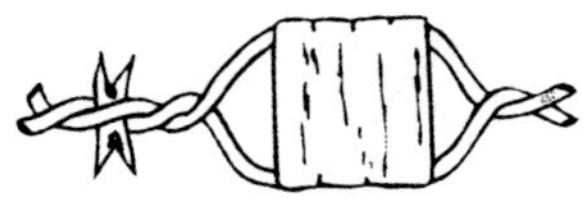

Wooden Block
H.B. Scutt 1880

Crandal's Champion
E.M. Crandal 1887

Stubbe Plate
John Stubbe 1883

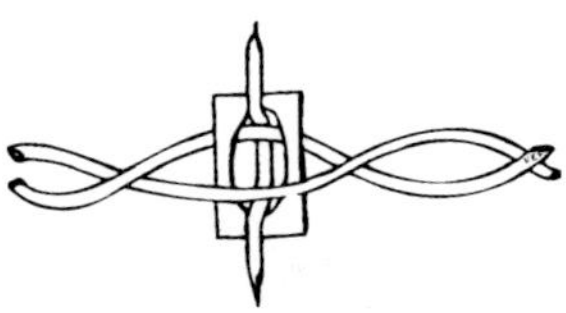

Clinch Wire
C.A. Vosburgh 1876

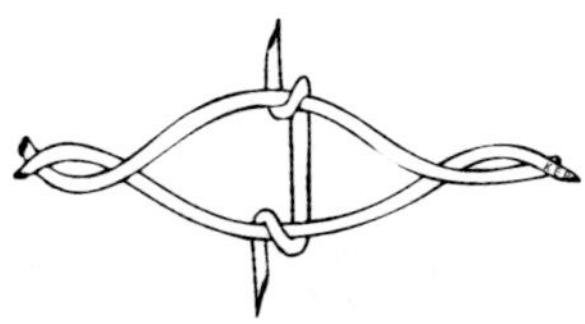

Decker Wide Spread
A.C. Decker 1884

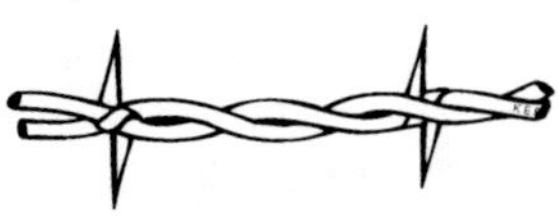

Baker Perfect
G.C. Baker 1883

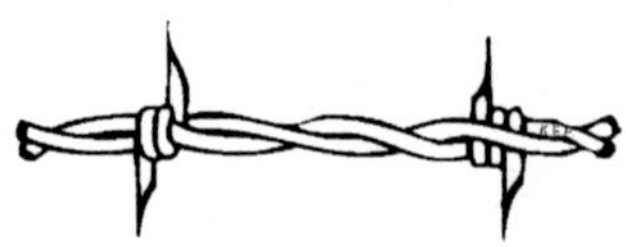

Glidden's Winner
J.F. Glidden 1874

TIMELINE REFERENCE GUIDE
1859 - 1897

EVENT	DATE
Boulder Canyon Gold Rush, thousands of gold seekers rush to the new gold fields in Colorado.	1859
Virginia City, Nevada, population explodes to the recent announcement of a silver discovery at the Comstock Lode.	
Cattle drives into British Columbia from southern routes begin.	
Two and a half million people immigrate to the United States over a period of ten years.	
Pony Express begins mail run, almost 2,000 miles long.	1860
30,000 settlers leave Kansas over severe drought conditions.	
Abraham Lincoln inaugurated as the the 16th President of the United States.	1861
Lincoln opposes slavery, South Carolina secedes from the Union.	
Texas forces Governor Sam Houston out of office and joins the Confederacy.	

American Civil War begins.

First transcontinental telegraph line
in service coast to coast.

Kansas pro-Union guerrillas cross into Missouri, killing citizens and burning most of the city of Independence.

US President Lincoln signs into law 1862
the Homestead Act.

The Five Civilized Tribes are divided by the Civil War over slaves brought with them them from the South when they were forced on the long trek to Indian Territory. (Trail of Tears)

Confederate guerrillas under the leadership 1863
of William Clarke Quantrill operate out of Missouri. Among the Quantrill Raiders were Frank and Jesse James and the Younger Brothers.

American Civil War over, surrender documents 1865
are signed by the Confederates at Appomattox Courthouse, Virginia.

Thousands of Civil War veterans take advantage of the Homestead Act.

Andrew Johnson takes office as the 17th President of the Unites States.

General P.H. Sheridan takes charge of western forces on the frontier in an attempt to bring peace by killing all the buffalo on the plains. Slaughter the buffalo and you eliminate the Indians. 1866

Frank and Jesse James rob bank in Liberty, Missouri.

First cattle trail from Texas to New Mexico established by Goodnight and Loving.

First cattle drive starts up the Chisholm Trail towards the railyards in Abilene, Kansas. 1867

First Patent issued for Barbed Wire to Lucian B. Smith, June 25th, Kent, Ohio.

Medicine Lodge Treaty enacted in order to remove Indians from the path of white settlement.

Barbed wire patent issued to Michael Kelly. 1868

Central Pacific and Union Pacific Railways meet at Promontory Point northeast of Ogden, Utah, linking both railways and forming a coast to coast rail system. 1869

Ulysses S. Grant becomes the 18th President of the United States.

Women get the right to vote in Wyoming.

Huge advertising campaigns by the railway companies to entice new settlers to the west. 1870

Utah grants women the right to vote.

School segregation begins in California.

Indian Appropriation Act ends Indian sovereignty and Indians are treated as wards of the government. 1871

Apache Chief Cochise, who terrorized Arizona white settlers for ten years, surrenders to the authorities.

Two Million acres set aside in Wyoming under the Yellowstone Act. 1872

Congressional Medal of Honor is awarded to Buffalo Bill (Cody) for service in the Army as Scout during a four year battle against the Cheyenne.

Cable cars in operation in San Francisco. 1873

Government estimate that three million buffalo are killed each year on the western plains.

Barbed wire patent issued to Henry M. Rose.

Joseph F. Glidden awarded patent for Barbed 1874
Wire called Glidden "Winner" on November 24th.

10,000 lbs. of barbed wire made and sold.

Barbed wire patents issued to Jacob Haish, Isaac Ellwood, Charles Kennedy and several more.

Gunfighter Wild Bill Hickock murdered in 1875
the Black Hills town of Deadwood.

5,000 Miners intrude into the Black Hills of 1876
Dakota for gold, a direct violation of the Lakota Treaty.

Custer's Last Stand against Chief Sitting Bull at Little Big Horn, Montana. His troops are slaughtered, all are lost.

Sitting Bull and his band escape the American Army and cross over into Canada.

2,840,000 lbs. of barbed wire made and sold.

John W. Gates ("Bet-A-Million") stages a daring demonstration in the town of San Antonio, Texas. He built a corral made of post and barbed wire, placing high spirited longhorns inside. The cattle were then stampeded, the fence held, apart from a few posts being broken. Not one animal escaped. Thousands of lbs. of wire was sold on the spot.

Barbed wire patents issued to E.D. Crandal, Hiram B. Scutt, Joseph F. Glidden, Andrew J. Upham, Jacob Brinkerhoff, John C. Merrill and many, many more.

Desert Land Act allows settlers to purchase 1877
640 acres of public land in arid regions where more land is required for large farming operations.

Congress repeals the 1868 Fort Laramie Treaty and returns the Black Hills to the Union, as well as an additional forty million acres from the Lakota.

Texas gunfighter John Wesley Hardin is jailed in a Texas State Prison for twenty five years for murdering a deputy sheriff.

Post Civil War Reconstruction era ends.

19th President of the United States, Rutherford B. Hayes.

12,836,000 lbs. Of barbed wire made and sold.

Barbed wire patents issued to Alexander C. Decker, George G. Hunt, Arthur S. Burnell, Frank Randall and many more.

Forty thousand African Americans immigrate to Kansas and are known as the Exodusters. 1878

26,655,000 lbs. Of barbed wire made and sold.

Barbed wire patents issued to John McNeill, Hiram Reynolds, Charles Washburn, Thomas Dodge, Ole O. Kittleson, Jacob Brotherton and many more.

Kansas passes new law and becomes the first state to "Go Dry," thereby ending the drinking saloons in towns like Dodge City. 1880

80,500,000 lbs. Of barbed wire made and sold.

Barbed wire patents issued to Thomas G. Orwig, Jacob Stoll, Thomas Shuman and many more.

Last big cattle drive to Dodge City ends 1881

Increasing use of Barbed Wire effectively ends the Open Range era.

President Elect, James A. Garfield, Inaugurated 20th President of the United States.

Chester A. Arthur becomes 21st President of the United States

Deputy Marshall Wyatt Earp and his brothers of Tombstone, Arizona gun down the Clantons at the OK Corral.

Sitting Bull returns from Canada and turns himself over to General Terry.

Billy the Kid is trailed by Sheriff Pat Garrett for more than six months before being killed by him at Fort Sumner, New Mexico. Billy had been charged with more than twenty murders.

120,000,000 lbs. of barbed wire made and sold.

Barbed wire patents issued to Milto Shinn, Caleb Brinton, C.K.Marshall, Ivu Wright and many, many more.

Chinese immigration falls from 40,000 in 1882
1881 to a mere 23 in 1882.

Outlaw Jesse James is killed with a shot in the back by a relative in hopes of collecting a $5,000 reward.

160,000,000 lbs. of barbed wire made and sold.

Barbed wire patents issued to O.P. Briggs, Scott Smallwood, N.Clark, W.E.Brock, J. Carpenter and many others.

The last large buffalo kill on the northern 1883
plains ends.

Buffalo Bill (William Cody) promotes his Wild West Entertainment Show.

Half of Texas is reporting mass cutting of barbed wire fences and the burning of range grass. Damages were estimated to exceed twenty million dollars.

200,000,000 lbs. of barbed wire made and sold.

Barbed wire patents issued to at least 45 inventors.

Cattle Industry in the mid west in trouble due to overgrazing. 1886

Two hundred mile Drift Fences constructed to help save the winter grasses.

Winter storms and Drift Fences are responsible for the death of cattle in the thousands, a period called the big or great "Die-Up".

Apache Chief Geronimo surrenders to General Miles in Skeleton Canyon, Arizona, ending a decade of warfare. Geronimo and his Tribe are exiled to Florida.

P.W. Putnam receives patent for the first power operated machine for applying barbs.

300,000,000 lbs. of barbed wire made and sold.

Barbed wire patents issued to W.H. Perry, J. Hunt, H.C.Pratt, William C. Gholson and several others.

The age of the Industrial Revolution begins, transformation of the west is for the most part complete. 1887

120,000 newcomers arrive on the western frontier by trains from the east.

320,000,000 lbs. of barbed wire made and sold

Barbed wire patents issued to less than ten new inventors.

At noon on April 22 a single gun shot signals white settlers to begin the lawful race to stake and claim land. In less than twelve hours, the Oklahoma Land Rush claims nearly two million acres of tribal land. 1889

23rd President of the United States, Benjamin Harrison.

370,000,000 lbs. of barbed wire made and sold.

Barbed wire patents issued to Merritt B. Chappell, John Schlyer, Orlando Huffman and several others.

By 1889 hundreds of patents for barbed wire were issued.

Johnson County War, Wyoming. In total state of lawless conflict between the big cattle ranchers, farmers and small cattle owners. Cattle rustling and illegal branding of cattle by both side, each side hopelessly locked in a violent effort to exercise their rights as they saw them. 1892

American Automobile Industry is born. 1896

BIBLIOGRAPHY

BIBLIOGRAPHY

Henry D. & Frances T. McCallum - The Wire That Fenced The West. - University of Oklahoma Press : 1965

Jack Glover - The Bobbed Wire Bible - Cow Puddle Press : 1996

Robert T. Clifton - Barbs, Prongs, Points, Prickers & Stickers - University of Oklahoma Press : 1970

Robert O.Campbell & Vernon L.Allison - Barriers - Western Profiles Publishing Company : 1986

Delbert and Ruth Trew - Notes and Tales on the History of Barbed Wire Fencing - Bull Canyon Press

Charlie and Rosie Dalton - The Barbed Wire Collector, Pocket Book of Wires : 1995

Art Edwards - Unusual Fence Wires : 1979

Dean Grauberger - Happiness Is Being A Barbed Wire Collector

Newsletters - New Mexico Barbed Wire Collectors Association

Newsletters - Wire Collector News - American Barbed Wire Collector's Society - John Mantz, (founder)

The Barbed Wire Collector - D.R.M. Publishing Company

Stan McLean - History of the O'Keefe Ranch - Unisco Press : 1984

Ralph Hall - Pioneer Gold Seekers of the Omineca - Morris Publishing : 1994

John Mantz - National Barbed Wire Collectors Membership Directory : 1994

John Mantz - GET "HOOKED" ON BARBED WIRE! : (Copyright 1997 by John Mantz)

Virginia Paul - This was CATTLE RANCHING Yesterday And Today - Superior Publishing Company Seattle, Washington : 1973

LANDMARKS and BRANDING IRONS, A Guide to some Historical Ranches in the Nicola Valley of British Columbia, Canada : 1988

Lewis Atherton - THE CATTLE KINGS - Indiana university press : 1961 - Bison Book Edition

I.L. Ellwood & Company - POCKET COMPENDIUM - Chicago : Press of Cameron, Amberg & Co. : 1887

Baker Wire Company - MEMORANDUM AND REFERENCE FOR 1891- Chicago: Press of Cameron, Amberg & Co. : 1891

Geoffrey C. Ward - THE WEST An Illustrated History: 1996 - Little, Brown & Company (Canada) Limited.

Richard A. Wright - DETROIT, Inc.: Wayne State University, Department of Communications.: 1996

LENOX TWIST
G.S. LENOX 1883

REYNOLDS NECKTIE
H. REYNOLDS 1878

A MESSAGE TO THE READER

Now that you have reached the end of this book, I hope it was informative, educational, and has enlightened you to the realm of barbed wire.

This book, of course, is by no means an exhaustive work on the subject of barbed wire or its historical significance. It was intended to introduce some basic facts and increase awareness of a product which is often taken for granted as "just another fence." It was also an attempt to relate a special story on the role that barbed wire played during the 1800's, and in essence the impact it had on the lives of so many during the settlement of North America's last frontier, as well as the lasting effect it has had in no small measure on our lives in these times.

I encourage everyone to further their knowledge of this most interesting subject, which is rich in history and steeped in intrigue. For a more comprehensive explanation of barbed wire and the people behind the inventions and manufacturing plants, I recommend reading the books listed in the Bibliography.

I also recommend and encourage those interested in this subject to start your own wire collection; you won't be disappointed. The hobby of wire collecting is enjoyed by thousands of men, women and children of all ages. Starting your collection is easy and enjoyable. Go out into the back country and see for yourself just how many kinds of wire there are; you will be surprised. Besides, it's a great way to get exercise, meet new friends, and experience the many wonders of nature.

Ken. E. Perry,
Author and Collector

Start Now!

GET "HOOKED" ON BARBED WIRE!

A NOTE FROM GOSS PUBLISHING:

Once you have decided to start your wire collection, you may want to purchase our "Barbed Wire Log Book," available from Goss Publishing at the address on the back cover or at the store where you purchased this book.